STARK LIBRARY   JUL -- 2020

DISCARD

D1716844

Animal Defense!

CHEMICAL COVER
Smells and Poisons

Enslow Publishing
101 W. 23rd Street
Suite 240
New York, NY 10011
USA
enslow.com

Emma Carlson Berne
and Susan K. Mitchell

These books are dedicated to Emily, who inspired the author.
—Susan K. Mitchell

Published in 2020 by Enslow Publishing, LLC.
101 W. 23rd Street, Suite 240, New York, NY 10011

Copyright © 2020 by Enslow Publishing, LLC.

All rights reserved.

No part of this book may be reproduced by any means without the written permission of the publisher.

Library of Congress Cataloging-in-Publication Data

Names: Berne, Emma Carlson, author. | Mitchell, Susan K., author.
Title: Chemical cover : smells and poisons / Emma Carlson Berne and Susan K. Mitchell.
Description: New York : Enslow Publishing, 2020. | Series: Animal defense! | Audience: Grade 3–6. | Includes bibliographical references and index.
Identifiers: LCCN 2018047378| ISBN 9781978507159 (library bound) | ISBN 9781978508095 (paperback)
Subjects: LCSH: Animal chemical defenses—Juvenile literature. | Animal defenses—Juvenile literature.
Classification: LCC QL759 .B47 2020 | DDC 591.47—dc23
LC record available at https://lccn.loc.gov/2018047378

Printed in the United States of America

To Our Readers: We have done our best to make sure all website addresses in this book were active and appropriate when we went to press. However, the author and the publisher have no control over and assume no liability for the material available on those websites or on any websites they may link to. Any comments or suggestions can be sent by email to customerservice@enslow.com.

Portions of this book originally appeared in *Animal Chemical Combat: Poisons, Smells, and Slime.*

Photo Credits: Cover, p. 1 (poison arrow frog) Dirk Ercken/Shutterstock.com; cover, p. 1 (green skin) Sebestyen Balint/Shutterstock.com; p. 5 © iStockphoto.com/Click48; p. 10 Elliotte Rusty Harold/Shutterstock.com; p. 13 Debbie Steinhausser/Shutterstock.com; p. 15 Rapeepong Phataiwunnakul/Shutterstock.com; p. 17 critterbiz/Shutterstock.com; p. 20 Agnieszka Bacal/Shutterstock.com; p. 22 Hoang Mai Thach/Shutterstock.com; p. 25 Klaus Ulrich Mueller/Shutterstock.com; p. 26 Thorsten Spoerlein/Shutterstock.com; p. 30 Bildagentur Zoonar GmbH/Shutterstock.com; p. 33 Kasira Suda/Shutterstock.com; p. 38 Vader Chen/Shutterstock.com; p. 39 Mark Moffett/Minden Pictures/Newscom; p. 41 David Wrobel/Visuals Unlimited/Getty Images; p. 43 Henry Guttmann/Hulton Archive/Getty Images.

Contents

Introduction 4

1. Ooze, Spray, Goo, and Squirts 8

2. Skunk Ahead! 16

3. Poisonous Little Pretties 24

4. A Beetle with a Bomb 32

5. Slime from the Deep 40

Glossary 46

Further Reading 47

Index 48

Introduction

A girl and her dog are hiking through scrubby bushes near the Florida shoreline. Suddenly, a large yellowish toad hops in front of her. It's about the size of a softball. Over its shoulders are large pouch-like swellings. The dog barks at the big toad and tries to bite it. She urgently calls off her dog!

Her pet is about to get poisoned by a cane toad! Cane toads ooze a deadly poison from the swellings over their shoulders, called paratoid **glands**. The poison hurts the heart and can kill dogs that bite cane toads. Humans have even died after eating cooked cane toads.

Cane toads are one of many animals that use chemicals to defend themselves. Puffer fish, for instance, contain a deadly toxin known as tetradotoxin. One

Introduction

A cane toad oozes white poison from its paratoid glands. The scientific name for the poison is bufotoxin.

puffer fish has enough tetradotoxin in its body to kill thirty people—and there is no cure.

Some animals have chemical defenses that are merely unpleasant rather than deadly. The shrew, a small rodent, gives off a bad smell when it is threatened by a predator. This usually makes the predator back off. Predators don't

CHEMICAL COVER: Smells and Poisons

like bad smells any more than humans do! These animals are considered by scientists to be "poisonous."

But what about rattlesnakes, bees, or the spiny lionfish? Aren't *they* poisonous? After all, they have dangerous, deadly chemicals in their bodies, too. No, these animals are not considered poisonous, scientists say. Instead, they are considered "venomous." The difference is that while all of these animals use chemicals to defend themselves, the venomous animals need to enter the predator's body. Rattlenakes have fangs. Bees have stingers. Lionfish have sharp spines. Venomous animals use sharp body parts to pierce the skin of the predator animal and deliver the toxins that way. Animals that are merely poisonous squirt or spray or ooze their poison instead.

These chemicals are sometimes made in the animal's body. Sometimes, they get these chemicals from eating certain foods. Others use poisons to make themselves taste terrible to eat, so predators will stay away. Slimy goo, milky ooze, sprays, streams of liquid, and even gases: chemical defenses take on all these forms.

Introduction

And chemical defenses are popular in the animal kingdom. Mammals, **amphibians**, insects, and fish use smelly, bad-tasting, or poisonous chemicals to survive. Even a few birds are poisonous. One is the hooded pitohui (PIT-oo-wee). The other is called the ifrita. Both birds live in New Guinea. Their feathers and skin are very **toxic**. Scientists believe that some of the things they eat make them poisonous.

Other birds use chemical weapons, but their bodies are not toxic. And some have weird ways of delivering their defensive chemicals. The fulmar is one of these. When it is threatened, this bird vomits an oily chemical at its attacker. This oil can seriously damage the feathers of other birds. The fulmar can launch its gross weapon up to 5 feet (1½ meters)!

Animals are in a constant battle: eat or be eaten. Prey animals must always try to defend themselves against predators. Some animals have stingers, some have armored skin, and some hide themselves using camouflage. Chemical poisons can seem unpleasant and a little scary to people. But they are just another method animals use to stay alive.

Chapter 1

Ooze, Spray, Goo, and Squirts

When someone hears the word "chemicals," they probably think of cleaning spray or bleach. Or maybe they think of factories churning out liquids to be used in manufacturing. These things are all chemicals. But nature makes chemicals, too.

Chemicals are found everywhere. They are in plants, water, and air. They are also in animals. Most chemicals used by animals are harmless. But a few of them can be very dangerous. Some chemicals are smelly. Some are just plain sticky. And then there are those deadly ones. These chemical defenses can often be very strange. But no matter how weird they may seem, chemical defenses can protect animals from being eaten.

Ooze, Spray, Goo, and Squirts

Mmm! A Poisonous Meal!

There are two ways that an animal can get the chemicals it needs for defense. The first way is through the food it eats. By eating plants or insects that contain poison, some animals become toxic themselves. The monarch butterfly is a good example. It has no way to produce its own poison. Its caterpillars are not born toxic. They get that way only after they start eating milkweed.

The milkweed plant is poisonous to many animals. Monarch caterpillars are able to eat it safely, however. In fact, it is the only food that a monarch caterpillar eats. The poisons of the milkweed plant stay stored inside the caterpillar's body.

When the caterpillar changes into a monarch butterfly, these chemicals make the butterfly taste terrible. Any predator that tries to take a bite of the monarch butterfly will become very sick.

The second way that an animal can have a chemical defense is to make the chemical inside its own body. All animals make chemicals, which are used simply to keep their bodies working. Some chemicals digest food.

CHEMICAL COVER: Smells and Poisons

A brilliantly colored monarch caterpillar eats milkweed. The caterpillar will grow into a beautiful but toxic butterfly.

Others carry oxygen through the body. Most of these chemicals are not dangerous at all. But some animals are able to make dangerous ones.

These poisons are produced by glands in the animal's body. They are stored in the glands until the animal uses them in an attack. For the unfortunate

Ooze, Spray, Goo, and Squirts

predator who comes across them, most poisons simply cause itching or burning. But others can kill.

Using chemical weapons can be pretty tricky. To use their weapons, animals often have to get close to a predator. This, of course, is very dangerous. So each

Get Ready for Gross: Vultures and Vomit

The vulture has a particularly smelly weapon at its disposal. When it feels threatened by a predator, it can vomit. Since vultures are scavengers and feed on dead animals, one can imagine that a puked-up already-dead raccoon smells pretty bad. Most predators will run away when they smell the horrible vomit. But just in case they need some more encouragement to leave, the fumes from the vomit will also sting their eyes and faces. Plus, the vulture has an empty stomach after vomiting. The bird is lighter, so it's even easier to fly away from the predator. It's a win-win for the vulture!

CHEMICAL COVER: Smells and Poisons

animal that makes its own chemical weapon has to have a good way to deliver it. Some have body parts that spray chemicals at their enemies. In others, the chemicals ooze from the skin. No matter how an animal delivers its chemical defense, the goal is to get away alive.

Not all animal chemicals are dangerous. Some are more like a message system. One important chemical message is sent by **pheromones** (FAIR-a-mones). Pheromones are made by an animal's own body. Other animals can detect pheromones by their smell. These chemicals work like a kind of language for animals. They are used to send many different types of messages, which are picked up only by other animals.

> A tiger leaves its scent on a tree by rubbing its face against it and scratching it. The pheromones that come from scent glands on the tiger's face and paws mark its territory.

CHEMICAL COVER: Smells and Poisons

Pheromones can help an animal find a mate. They can warn other animals when predators are close. Some pheromones can also mark an animal's living and hunting area. Animals often rub against trees or bushes where they live. They might also spray objects with fluids from their bodies. This rubbing and spraying leaves tiny bits of scent behind. By leaving their scent with pheromones, animals warn others to stay away.

Fun Fact!

Possums try to fool predators by pretending to be dead. But if that doesn't work, they can poop out a bad-smelling green mucus to make predators think again!

Ooze, Spray, Goo, and Squirts

This colorful cuttlefish is signaling "Watch out! I'm poisonous!" Cuttlefish store a deadly poison in their muscles.

Look Out! I'm Poisonous!

Many animals need to hide from predators using camouflage. They often have dull colors to help them blend in with nature. Animals with chemical defenses do not have to worry about that. They often move around in plain sight. Many of them advertise their chemical defenses with skin that is brightly colored.

These bright colors warn predators. Over time, predators have learned that these animals are not good to eat. It only takes a few match-ups with a stinky, bad tasting, or slimy animal to teach predators a lesson.

Chapter 2

Skunk Ahead!

Skunks are some of the best-known smelly animals. They often live near people—and sometimes people or dogs get a little *too* close. Skunks are not large, but they are powerful. They can shoot an extremely smelly chemical from their backside straight at a predator, such as a dog, that might be threatening them.

The great horned owl is the skunk's only regular predator. The owl can swoop down on the skunk so quickly, the skunk doesn't sense the owl until it is too late. In addition, the great horned owl has a bad sense of smell, so it tends not to care if it does get sprayed.

There are many different kinds of skunks. They can be striped or spotted. Some skunks, such as the hog-nosed skunk, have an all-white back. Their coloring signals to predators to leave them alone.

Skunk Ahead!

But skunks do not just have their own chemical defenses. They can also survive chemical attacks dished out by other animals. For example, a skunk can often survive a snakebite that would kill another animal.

Skunks lift their tails when they are ready to spray. Skunk spray can temporarily blind animals if they get the chemical in their eyes.

CHEMICAL COVER: Smells and Poisons

A skunk's chemical weapon is located on its rear end. There are two tiny holes below the base of the skunk's tail. Inside each hole is a tiny scent gland. The scent glands are full of a stinky, oily chemical. When a skunk is threatened, it turns, aims, and fires! It squeezes tiny muscles that cause a jet of yellow oil to squirt out of the holes. That oil is super stinky. Skunks can carefully

Here, Boy! Skunks as Pets

It is illegal to keep skunks as pets in some states. But where it is allowed, people have their pet skunks denatured, which means they have the skunk's scent glands removed. Then the skunks can't spray. But skunks are not used to living with humans, like dogs and cats. They are nocturnal, so they are awake when people are sleeping. They dig all the time and will rip up furniture. And they get bored and frustrated living in a house. Skunks need the natural activity of living outdoors, in the wild, where they belong.

control their squirting. They can spray a fine mist or a thick jet stream.

A skunk tries to aim at the face of a predator. That is because the chemical is not just stinky. It stings and burns. This stops the predator long enough for the skunk to run away and hide!

A Shy Creature

Skunks use their secret chemical weapon only as a last resort. They are actually very shy animals. Most skunks live in burrows or dens. Skunks spend most of the day hiding and resting. They are usually nocturnal, which means they are active at night. After the sun goes down, skunks come out to hunt for food.

Since skunks do not see very well, they use their nose to find food. Skunks sniff along looking for hidden bugs and other treats. They have been known to sniff out food that is several inches underground.

Skunk Gymnastics

During daytime naps or nighttime hunts, a skunk can often run into trouble. Large predators such as coyotes or

CHEMICAL COVER: Smells and Poisons

An eastern spotted skunk does a handstand to ward off a predator. Its bright white spots tell the predator, "Stay away from me!"

bobcats may get curious and stop to check out a skunk. If the skunk feels threatened, it has several tricks to use before it fires a stink bomb.

The first thing a skunk will do is try to **bluff** the predator. This means the skunk puts on a show to look and sound meaner than it really is. Skunks raise their tails high and puff out their fur. They stamp their feet while they growl and hiss.

Skunk Ahead!

If the bluff does not work, the spotted skunk will even do a handstand! It stands on its front paws with its back legs and tail high in the air. This helps it look larger. It also helps the skunk get a good aim if the bluff is not working. If the predator does not back off after all this, it is time to use the secret chemical weapon.

Whether the skunk is standing on its front paws or on all fours, it first turns its backside to the enemy. Then it fires its stinky stream of skunk oil.

People find skunk odor hard to wash off of their dogs, possessions, or themselves. Since the skunk's chemical is an oil and oil and water do not mix, water will not wash it away. Some people say that using baking soda and dish detergent is the best way to get rid of the smell.

Slow and Smelly

There aren't many poisonous mammals, but the slow loris is one. This animal is closely related to monkeys and apes. Like its name says, the slow loris is a

CHEMICAL COVER: Smells and Poisons

The slow loris may look cute and gentle, but the poison it produces can harm, or even kill, some humans.

Skunk Ahead!

Fun Fact!

After a skunk sprays and empties its glands, it takes ten days before they fill up again.

very slow-moving animal. It lives in trees. It also has giant eyes. These big eyes help the slow loris hunt for insects at night.

On the inside of the slow loris's elbow is a small gland. It produces a poison that smells like stinky gym socks. The slow loris uses the poison in a couple of ways. It can rub the poison on its teeth so it will have a toxic bite. The slow loris also rubs the smelly toxin on a baby loris's fur. This makes the baby loris instantly poisonous. Then the baby is protected from predators while the parents are out hunting.

Chapter 3
Poisonous Little Pretties

Clinging to a leaf in the dark, green tropical rain forest in South America is a frog that looks more like a jewel and is no bigger than a nickel. Red and black, it glows against the bright green. Another one is blue and black. And yet another frog is gold and black. These are poison dart frogs. They are both beautiful *and* dangerous. Their bright colors warn predators to watch out!

Even the largest poison dart frog is only a couple of inches long. These little frogs pack quite a punch, however. Inside their skin are poison glands. This poisonous skin is how the frogs got their name. Native tribes who live in the jungles often rubbed the tips

Poisonous Little Pretties

A blue poison dart frog shows off its stunning colors. These toxic little frogs also come in a variety of other colors, including red, copper, gold, green, and black.

of their darts on the frog's skin. This coated their darts with poison. They used the deadly darts to hunt for food.

Some poison dart frogs have poison that just tastes bad to predators. Others have poison that causes burning and stinging when it is eaten. And there are a few poison dart frogs that are truly deadly. Some have a poison so strong that less than two micrograms of it could kill any predator. That is less poison than would fit in the period at the end of a sentence. One

CHEMICAL COVER: Smells and Poisons

particularly deadly frog is the golden poison dart frog. This bright golden-yellow frog has a strong enough poison in its skin to kill a human.

Poison dart frogs have only one known predator. It is a snake with the scientific name *Leimadophis epinephelus*. A poison dart frog's toxin will usually not kill the snake. But if the toxin is strong enough, the snake could still get a little sick.

Poison dart frogs are not born poisonous. The poison comes mostly from the food they eat. Scientists think that it takes a whole "chain" of food to give the poison dart frog its poison. For example,

The amount of toxin in one golden poison dart frog could kill up to ten grown men.

Poisonous Little Pretties

Fun Fact!

Poisonous frogs don't just live in the rain forest. The gray tree frog makes its home in the southeastern United States and oozes a toxic liquid from its skin.

insects eat plants that produce toxic substances. The plant's poison is then in the insect. Then the poison dart frogs eat those insects. The poison dart frog's body allows it to eat the poison without being harmed. The poison then becomes part of the poison dart frog's body chemistry.

A poison dart frog that is taken from the jungle and fed regular insects will become much less poisonous. In fact, poison dart frogs that are born in captivity (not born in the wild) never have any poison at all.

CHEMICAL COVER: Smells and Poisons

No Hide and Seek

Poison dart frogs are diurnal. This means they are active during the day. The frogs' bright colors keep them safe from predators while they are busy hunting ants and other insects in daylight.

Poison dart frogs are amphibians; they live both in water and on land. These tiny frogs place their babies, called tadpoles, inside the bromeliad (bro-MEE-lee-ad) plants that grow in the trees.

There, the poisonous parents bring small insects for their tadpoles to eat. Soon the tadpoles begin to change. They lose their tails and their skin starts to turn bright colors. Thanks to the insects that they ate, the skin of the young poison dart frogs is also now poisonous.

Salamanders, Frogs, and Poison, Oh My!

Poison dart frogs are not the only amphibians that have poisonous bodies. Salamanders can be just as toxic. The fire salamander is one of them. The small insects and worms these salamanders eat make them toxic.

Poisonous Little Pretties

If it is threatened by a predator, a fire salamander can spray its poison if it has time. But even if the predator is already too close and chomps down on the fire salamander, it is still in for a rude surprise.

Bad News for Poison Dart Frogs

Scientists already know that amphibians, such as frogs and salamanders, are some of the most threatened groups of vertebrates. But recently, researchers found that toxic amphibians, such as the poison dart frog, are at an even higher risk of disappearing from existence—of going extinct. Poison is good for defending against predators, the researchers concluded, but somehow, it is bad for continuing the species. Scientists are still trying to figure out why. One possibility is that chemical defenses might take more of the amphibian's energy than in a non-poisonous amphibian. Then the animal might be more likely to get sick.

CHEMICAL COVER: Smells and Poisons

Most of the salamander's poison glands are around its head and neck. Predators usually try to grab their food by the head. When a predator does this to a fire salamander, it gets a mouth full of poison. The burning, bad taste of the poison is usually enough to make the predator let go. This gives the fire salamander time to get away.

A fire salamander climbs a tree. This amphibian lives across wide swaths of Europe, northern Africa, and parts of Asia.

Poisonous Little Pretties

Some types of frogs produce even weirder things than poison. The tomato frog lives only in Madagascar. It can produce a sticky liquid from its skin when it is attacked. The liquid is white and looks very much like glue. It also acts like glue. The glue is not poisonous. But this stuff is so strong that it can glue a predator's mouth shut for several days!

Scientists are experimenting with the glue of this and other oozing frogs. They hope to someday be able to help repair human joints using the sticky stuff. They may also be able to repair human cartilage, which is the material that humans' noses and ears are made of.

The skin of any amphibian is amazing. It helps the animal breathe. It can also produce chemicals used for defense. The bad news is, a frog's skin is so sensitive, it can absorb chemicals that may be in the water or air. This can cause the frogs to become sick or die. Chemical pollution can also cause frogs to have fewer babies. By studying poison dart frogs, scientists can tell a lot about what is going on in the rain forest. Even tiny changes in the environment will affect the frogs before most other animals.

Chapter 4

A Beetle with a Bomb

The bombardier has an amazing chemical weapon inside its body. When it is threatened, the bombardier beetle mixes two chemicals together and shoots them out of its body. The chemicals explode in the air! And predators run. The bombardier beetle is tiny but mighty.

A bombardier beetle is built like other insects. It has three body parts: a head, a thorax, and an abdomen. The abdomen of the bombardier beetle is where it keeps its secret weapon. Inside the beetle's abdomen, there are special cells. They store two different chemicals.

If the beetle is attacked, the cells release their chemicals into another area in the abdomen called a **reservoir**. One of the substances is a common household

A Beetle with a Bomb

The bombardier beetle is found in Europe, North and South America, Africa, and Australia. It lives in grasslands and forests, hiding under leaves and other ground cover.

chemical used to clean cuts and scratches, called hydrogen peroxide. The hydrogen peroxide and the other chemicals are mixed to make a dangerous weapon.

In an instant, the bombardier beetle shoots the mixed chemicals out of a small tube located behind its abdomen. This mixture is stinky and irritating. It is so hot that it actually burns when it hits its target. This spray can reach 212 degrees Fahrenheit (100 degrees Celsius). That is the temperature of boiling water! The heat and force

CHEMICAL COVER: Smells and Poisons

of the chemicals makes some of the liquid turn to gas. The gas looks like a puff of smoke that comes out with the liquid.

There is also a loud bang when the bombardier beetle fires its weapon. The noise alone is enough to scare most predators. But the bombardier beetle also has a great aim. Its abdomen can twist in almost any direction. This gives it dangerous accuracy when it shoots its disgusting spray at an enemy.

Fun Fact!

The Latin name for the bombardier beetle is *Brachinus crepitans*. Crepitans means "crackle." It refers to the explosive sound the chemicals make when they eject from the beetle's back end.

A Beetle with a Bomb

Dangerous but Slow-Moving

Bombardier beetles spend most of the day hiding under rotten logs or leaves on the ground. At night, they search for food. They have wings, but they do not work well for flying. This slow-moving beetle cannot even walk away very fast! It has almost no way of escaping a predator.

Ants are the biggest threats to the bombardier beetle. But no single ant can do the job. Instead, huge numbers of ants gang up to overwhelm their prey. Hundreds of ants can take apart a large insect in a matter of minutes. For the bombardier beetle, sharing the ground with these tiny hunters is dangerous business. The beetle needs a revved-up weapons system so it can stay and fight.

Key Chemical Reactions

The bombardier beetle also needs its great aim when fighting against ants. Ants can attack from any direction, so the beetle needs to be able to fire its chemicals in any direction. It also needs to be able to fire more than once. In fact, the bombardier beetle can fire its weapon many times in a matter of seconds.

CHEMICAL COVER: Smells and Poisons

The bombardier beetle's chemicals are hot enough to burn, and they explode out of the beetle with enough force to make a loud noise. So what keeps the bombardier beetle from blowing itself to bits? The answer is another chemical reaction.

Inside the reservoir in the beetle's abdomen where the chemicals mix, other cells release enzymes.

Exploding Vomit

Some frogs in Japan ate bombardier beetles. Then the bombardier beetles, which were still alive, released their explosive chemicals from within the frogs' stomachs. Ouch! The frogs vomited the beetles back up. The beetles came out of the frogs' mouths alive and well.

Scientists studying the frogs and the beetles discovered that the bombardier beetles were unusually tough. The frogs' digestive juices didn't bother the bombardier beetles very much, even after twenty minutes in the frogs' stomachs. The bombardier beetles might have evolved to survive being eaten by the frogs, the researchers believed.

A Beetle with a Bomb

Enzymes are made naturally by all living things. An enzyme's job is to help speed up chemical reactions. When the enzymes inside the bombardier beetle mix with the chemicals, heat and pressure build up quickly. The chemicals literally explode out of the beetle's rear end. This all happens so fast that the chemicals are released before the beetle can blow itself up.

Beyond the Bombardier Beetle

The bombardier beetle is not the only insect that survives using chemical defenses. A stinkbug's chemicals might not be as dramatic as the bombardier beetle's, but its weapon is effective! There are many kinds of stinkbugs. They can be many different sizes and colors. Most stinkbugs have a flattened body shaped like a shield. The one thing they all have in common is their chemical defense. If bothered, these insects can release a very smelly liquid. Few predators want to eat something that stinks.

The chemicals are kept in glands on the stinkbug's thorax. It does not take much to get a stinkbug to release

CHEMICAL COVER: Smells and Poisons

its bad odor. A simple bump or squeeze on its sides makes the glands react.

The exploding ants of Malaysia use their own bodies as deadly weapons to protect other ants in their colony. When these ants are threatened, special chemicals in their bodies mix together just like in the bombardier beetle. When that happens, the ants explode, spraying toxins all over the place! The ant dies, but its family survives.

Insects and related animals can provide chemical defense even when they don't intend to—like after they are dead! Some kinds of monkeys rub crushed millipedes on their fur. Millipedes are animals similar to

A yellow spotted stinkbug rests on a twig. Not only do stinkbugs smell bad, but they eat and damage farm crops, too.

A Beetle with a Bomb

The smaller brown ant has exploded its body to kill the larger black ant. The sticky yellow toxin is sprayed over both.

insects with many legs and long bodies made up of several segments.

Some millipedes are pretty toxic. So when monkeys rub them on themselves, scientists believe they are using the millipedes as a sort of insect repellent. The poisons of millipedes and other bugs help keep pesky mites and fleas out of the monkeys' fur.

Chapter 5

Slime from the Deep

An animal wriggles through the deep dark water near the ocean floor. This fish has a skull but no spine. It has no eyes or jaws, but it does have tentacles. It burrows into the flesh of dead or dying animals and eats them from the inside out. And it's covered in slime.

This animal is not the creation of a Hollywood special effects studio. This is the hagfish, a real ocean-dwelling fish. And despite its gruesome appearance and dining preferences, the hagfish is a fascinating animal that uses its chemical defense effectively. When a hagfish feels threatened, it creates a giant slime cloud around itself, and it can make huge amounts of the slippery stuff.

Slime from the Deep

The hagfish's appearance alone can be fairly frightening. Hagfish do not have scales like most fish. Their skin is smooth, like an eel's skin, and grayish-pink in color. They can grow to be around 2 feet (61 cm) long.

A Pacific hagfish rests on a sandy bottom. Since the hagfish lives very deep in the ocean where there is little light, it does not need well-developed eyes.

CHEMICAL COVER: Smells and Poisons

Fun Fact!

Hagfish don't have jaws but they do have two sets of hard structures in their mouths that are like teeth. They use these to burrow into the dead animals that they eat.

Hagfish do not have real eyes. They have no eye muscles or lenses. They may have only small round holes where the eyes should be. They are almost completely blind.

Eyes are not very important to hagfish, anyway. Hagfish have been known to live as deep as 4,000 feet (1,219 meters) below the ocean's surface. At those depths, it is very dark. The hagfish relies mostly on its great senses of smell and touch to find food.

Humans and Chemical Defenses

Since World War I (1914–1918), people have used chemicals, usually liquid or gas, during battle. Today, chemical weapons are often called weapons of mass destruction (WMDs) because they can kill or injure a large number of people at one time.

Early in the twentieth century, smell was usually the only way to know that dangerous chemicals were in the area. Today, chemical weapons may have no smell at all. Soldiers or emergency workers use lasers to find even the tiniest amounts of chemicals in the air.

German soldiers release poisonous gas from cylinders during World War I. The wind is blowing it away from them, which is why they aren't wearing gas masks.

CHEMICAL COVER: Smells and Poisons

Member of the Clean Plate Club

A hagfish eats mostly marine worms and other small ocean animals. However, it sometimes eats dead or dying fish from the inside out. That may be a nasty idea, but by doing this, hagfish help get rid of dead material on the ocean floor. This helps keep the water clean.

Many animals like to eat hagfish. Whales, seals, and some seabirds eat hagfish. But no fish with **gills** hunt for hagfish. This is because of the hagfish's slimy defenses. It has special glands all along its body that produce a sugary liquid if the hagfish is bothered. When this liquid hits salty seawater, it turns to thick slime. There are also tiny thread-like fibers inside the slime that help make it extra thick and gooey. A frightened hagfish makes slime by the bucketful! Any fish caught in this huge slime ball would soon be unable to breathe. The slime would coat its gills and the fish would suffocate.

The hagfish keeps from suffocating itself in a cloud of slime by twisting itself into a knot. It slides the knot down its entire body, wiping the slime away from the

Slime from the Deep

hagfish's gills. A hagfish also sneezes to get the slime out of its mouth, tiny eyes, and one small nostril.

Using Chemicals to Survive

The chemical world can be complex and pretty hard to understand. Maybe it is because so many chemicals can do so many different things. The chemicals used by animals for defense are some of the wildest. They may be stinging, sticky, slimy, stinky, or deadly, but all of these natural chemical weapons give the animals that use them an amazing defense.

Glossary

amphibian A cold-blooded animal, such as a frog or salamander, that can live both on land and in water.

bluff When an animal tricks another into thinking it is bigger, stronger, or scarier than it really is.

denature To take away something's natural qualities. Skunks are sometimes denatured, meaning their scent glands are removed.

digestive Relating to the stomach and the intestines, and to the system of digesting food.

evolve To change slowly over time.

extinct Not existing anymore.

gills The organs of a fish or other water-dwelling animal that allows it to breathe underwater.

gland A small structure in the body that produces a substance, such as venom or sweat.

pheromone A chemical produced by animals that gives off a scent and sends messages to other animals.

reservoir A part of the body that holds fluids.

toxic Poisonous.

vertebrate An animal with a spine. Mammals, reptiles, fish, birds, and amphibians are vertebrates. Invertebrate animals, such as insects, do not have spines.

Further Reading

Books

Cusick, Dawn. *Get the Scoop on Animal Snot, Spit & Slime!* Lake Forest, CA: MoonDance Press, 2016.

Higgins, Nadia. *Slimy Animals.* Minneapolis, MN: Jump!, 2016.

Hirsch, Rebecca E. *Exploding Ants and Other Amazing Defenses.* Minneapolis, MN: Lerner Classroom, 2017.

Johnson, Rebecca L. *Masters of Disguise: Amazing Animal Tricksters.* Minneapolis, MN: Millbrook Press, 2016.

Websites

National Geographic Kids: Poison Dart Frog
kids.nationalgeographic.com/animals/poison-dart-frog
Find out more about the poison dart frog.

National Geographic Kids: Skunks
kids.nationalgeographic.com/animals/skunk
Read interesting facts about this striped, smelly animal.

Neuroscience for Kids: Neurotoxins
faculty.washington.edu/chudler/toxin1.html
Learn about different animal poisons and how they affect the human nervous system.

Index

A
amphibians, 7, 28, 29, 31
ants, exploding, 38

B
birds, 7
bombardier beetles, 32–37

C
camouflage, 12, 14
cane toad, 4
chemical defense, 5, 6–7, 8, 9, 12, 15, 17, 29, 37, 38, 40, 43
chemicals, 4, 6, 7, 8, 9, 10, 12, 16, 18, 19, 21, 31, 32–33, 34, 35–36, 37, 38, 43, 45

G
glands, 4, 10, 18, 23, 24, 30, 37, 38, 44

H
hagfish, 40–45

M
millipedes, 38–39

P
pheromones, 12, 14
poison, 4, 6, 7, 9, 10, 11, 23, 24, 25, 26, 27, 29, 30 31, 39
poison dart frogs, 24–29
predator, 5–6, 7, 9, 11, 14, 15, 16, 19–20, 21, 23, 24, 25, 26, 28, 29, 30–31, 32, 34, 35, 37
puffer fish, 4–5

S
salamanders, 28–30
skunks, 16–23
stinkbug, 37

T
tetradotoxin, 4–5
tomato frog, 31

V
vertebrates, 29
vultures and vomit, 11

W
weapons of mass destruction (WMD), 43